A Robbie Reader

Mudslide in La Conchita, California, 2005

Karen Bush Gibson

P.O. Box 196
Hockessin, Delaware 19707
Visit us on the web: www.mitchelllane.com
Comments? email us:
mitchelllane@mitchelllane.com

Printing 1 2 3 4 5 6 7 8 9

A Robbie Reader/Natural Disasters

Earthquake in Loma Prieta, California, 1989
The Fury of Hurricane Andrew, 1992
Mt. Vesuvius and the Destruction of Pompeii, A.D. 79
Mudslide in La Conchita, California, 2005
Tsunami Disaster in Indonesia, 2004
Where Did All the Dinosaurs Go?

Library of Congress Cataloging-in-Publication Data
Gibson, Karen Bush.
Mudslide in La Conchita, California, 2005 / by Karen Gibson.
p. cm. — (Natural disasters—what can we learn?)
Includes bibliographical references and index.
ISBN 1-58415-418-7 (library bound)
1. Mudslide—California—La Conchita—Juvenile literature. I. Title. II. Series.
QE599.U5G53 2005
551.3'07'0979492—dc22
2005009698

ABOUT THE AUTHOR: Karen Bush Gibson has written extensively for the juvenile market. Her work has appeared in such publications as *Boys' Life* and *Cobblestone*. She is also the author of twenty-four school-library books. Karen lives in Norman, Oklahoma, with her husband and three children.

PHOTO CREDITS: Front cover, title page, pp. 4, 12, 15—Kevork Djansezian/AP Photo; title page, pp. 7, 11, 18, 20, 26, back cover—David McNew/Getty Images; p. 8—California Coastline; p. 10—Carlo Allegri/Getty Images; p. 16—Spencer Weiner/Getty Images; p. 21 (top)—Ric Francis/AP Photo; p. 21 (bottom)—Damian Dovarganes/AP Photo; p. 22—Ana Elisa Fuentes/Getty Images; p. 24—Rod Rolle/Getty Images.

PUBLISHER'S NOTE: The following story has been thoroughly researched and to the best of our knowledge, represents a true story. While every possible effort has been made to ensure accuracy, the publisher will not assume liability for damages caused by inaccuracies in the data, and makes no warranty on the accuracy of the information contained herein.

TABLE OF CONTENTS

Words in **bold** type can be found in the glossary.

Mudslides and flooding closed the roads in and out of La Conchita.

Rainy Days

People enjoy the many bright, sunny days that are common in Southern California. But there was no sunshine in Southern California on January 10, 2005. It was yet another day of heavy rainfall that had lasted for more than two weeks. The wet weather caused problems for the beachside community of La Conchita (lah kohn-CHEE-tuh). The roads south of town were flooded. A small mudslide had closed the highway north of town.

La Conchita resident Dena Hayess took pictures of the mudslide north of town. Then she returned home. She sat on her porch and talked on the telephone. Suddenly, the earth began shaking. She left the porch to look around the corner of her house. Much to her surprise, a house slid by.

Bill Harbison had recently moved to La Conchita. He decided to take his bicycle to see the storm damage. He looked up when he heard what sounded like a loud pop. Part of a mountain overlooking La Conchita had **collapsed** (kuh-LAPST). Thousands of **tons** of mud were speeding toward the town.

Brie Brazelton was at a store near her home when she heard rumbling like an **earthquake** (URTH-kwake). She left for home. But soon she had to change directions. A wall of mud was coming toward her. She ran for her life.

Diane Hart had just finished eating lunch when she looked out her kitchen window. She saw people running and screaming. She felt the rumbling and ran for a coat closet a few yards away. The closet was filled with pillows and blankets. She had waited in the closet earlier in the day when La Conchita was under a tornado warning. As she reached the closet, she saw the roof coming down. Her home started collapsing. Boards, rocks, and mud pushed at Diane Hart's closet from all sides. She couldn't get out.

Brie Brazleton (left) is comforted after the La Conchita mudslide. She ran from a wall of mud coming toward her.

La Conchita is a beachside town known for its surfing and views of the ocean. The yellow arrow is pointing to the retaining wall that was built to protect the town from mudslides.

History of California Mudslides

La Conchita is a small beachfront town of about 300 people in Ventura (ven-CHUR-uh) County, northwest of Los Angeles. Sitting between the Santa Ynez (SAN-ta YEH-nez) Mountains and the Pacific Ocean, it is a beautiful spot known for its views of the ocean and the surfing.

The people of La Conchita and other California towns are familiar with mudslides. On Christmas Day in 2003, mudslides in Devore, California, killed 12 people. La Conchita's history includes a 1909 mudslide that pushed onto the railroad tracks, taking four lives. In 1995, mud and dirt weighing 600,000 tons crashed down on the town. That slide

This campground playroom in Devore, California, filled up with mud during the 2003 Christmas Day mudslide.

destroyed four houses. Some people left La Conchita at that time. Others stayed. Ventura County spent $400,000 on a 280-foot **retaining** (rih-TAYN-ing) wall to protect the town from mudslides.

Scientists also placed **sensors** (SEN-surs) on the mountain behind La Conchita. Sensors would let people know about any movement on the mountain. Movement might mean a mudslide was about to happen.

Mud starts to slide when water builds up too quickly and weakens the soil. It can happen anywhere. California has had a large number of mudslides. The coast is lined with cliffs and bluffs that are weakened by **erosion** (ih-ROE-zhun). Wildfires destroy trees and other plants, which also weakens the soil. **Geologists** (jee-AH-luh-jists) look at how hard it is raining and for how long. If there has been a lot of rain plus movement on the mountain, they ask people to leave their homes for a safer place.

The end of 2004 and the beginning of 2005 brought very heavy rain to Southern California. Some places received more in two weeks than they normally get in a year. La Conchita had continuous rain, almost eight inches in five days.

Heavy rains started in California in late 2004 and continued into 2005. Vehicles like this car were getting stuck in mud pits.

This view from the top of the Santa Ynez Mountains shows the devastation caused by the mudslide. The mud poured right over the retaining wall.

An Unstoppable Wall of Mud

The heavy rains and closed roads on January 10 kept many of La Conchita's residents home from work. Many enjoyed having the day off. They watched television or took naps.

Weakened by the heavy rains, the cliff above the town suddenly cracked and crumbled. The sensors on the mountain gave no warning. A river of mud roared down the hillside. It pulled trees from the ground and broke them like toothpicks. The mud tore down power lines and carried them and all the rocks in its path. The retaining wall built to protect the town broke apart and became part of the **debris** (duh-BREE).

CHAPTER THREE

When the mudslide reached La Conchita, it was an unstoppable wall that swallowed up anything in its path. People who saw the mudslide ran as fast as they could away from it. Cars, houses, and people disappeared into the brown sea of mud. Houses were flattened like pancakes. A 17-year-old boy told his mother about seeing the mud smash people between cars. Screams and honking horns filled the air.

Jimmie Wallet had left his wife, Mechelle, and three youngest daughters (Paloma, age 2; Raven, 6; and Hannah, 10) at home while he went to get ice cream. His oldest daughter, 16-year-old Jasmine, was away at a friend's house. As he returned from the store, he saw the mud swallow up the house where his family waited for him. He yelled and ran to try to help them.

The mud covered four blocks of La Conchita. An unknown number of people were missing. Bill Harbison heard two women screaming. They were trapped under the mud. He immediately began digging. Soon other people joined him. They rescued the two women.

The mudslide ended in a 30-foot hill of mud. This is as high as a three-story house. National Public Radio reporter Carrie Kahn reported seeing "a mangled car **chassis** [CHAH-see], trees, drapes, and rugs" in the hill of mud.

The mudslide tore through homes, leaving behind tons of debris.

Rescue workers attend a memorial service for the victims of the 2005 mudslide.

Finding Survivors

Hundreds of rescue workers from the fire department, sheriff's department, and other agencies quickly arrived in La Conchita. The best guess was that the huge hill contained 15 houses and up to 21 people.

But where should they start digging? The rescuers listened for sounds. They dropped **microphones** (MY-kruh-fones) and cameras into the mud. Dogs trained to find people were brought on the scene. Emergency workers dug with shovels and with their hands. Meanwhile, other people watched the mountain. They held air horns to warn the emergency crews should another mudslide start.

Emergency workers search for survivors in the mud.

As rescue workers dug into the hill, they were happy to find small spaces with air where people could survive. Firefighter Joe Luna believed that people trapped in these air pockets could live for hours or even days until they could be found. And that's exactly how most of the people who were rescued managed to survive.

Greg Ray dived under two parked cars when the mudslide came. Searchers dug him out of his air pocket in three hours. Diane Hart was in another air pocket under 10 feet of mud. She had heard the people walking around. She was afraid she was going to die. She yelled to the searchers to let them know where she was. It took almost five hours to dig her out. She was taken to the hospital with broken ribs, broken **vertebrae** (VUR-tuh-bray), and a crushed left arm. Diane Hart was the last person to be dug out alive.

Rescue workers carefully dug with shovels, buckets, and hands. Their hands were chafed and bleeding after hours and hours of digging. Jimmie Wallet was worried that his family hadn't been found. He dug alongside the rescue workers. As Tuesday moved into Wednesday, six bodies were found. Early Wednesday morning, four more bodies were uncovered. They were Jimmie Wallet's wife and youngest daughters.

Jimmie Wallet holds toys that belonged to his three youngest children, who were killed in the mudslide.

After the mudslide, many homes, vehicles, and personal belongings were lost forever.

California governor Arnold Schwarzenegger visits La Conchita two days after the mudslide.

The Future of La Conchita

The early 2005 rains in California caused 10 deaths in other parts of the state, mainly because of flooding. The La Conchita mudslide killed 10 people and injured at least 14 others. Of the 166 homes in this coastal community, 15 were destroyed. County officials placed tags on 23 more houses. The tags showed that the houses were unsafe to live in.

California governor Arnold Schwarzenegger (SHWARTS-in-eh-gur) visited La Conchita two days after the mudslide. He declared Ventura County a disaster area. He also praised the townspeople for helping their neighbors. He told reporters, "In the past few days, we have seen the power of nature to

CHAPTER FIVE

Emergency workers look over the damage.

cause damage and despair, but we will match that power with our own resolve."

The Red Cross set up shelters for La Conchita families. The agency also kept track of information about the people of La Conchita. People from all over the country were calling to find out about friends and relatives. The names of missing persons were read at a **community** (kuh-MYOO-nuh-tee) meeting. It was a happy occasion when four of those people stood up to say they were there and they were fine.

Cleaning up and holding funerals occupied people in the days following the mudslide. Residents were allowed to look for personal belongings. Searchers continued digging, but they stopped when it became obvious that no one else was left alive.

The people of La Conchita had to make decisions about their futures. Geologists say the area is unsafe to live in. Many residents agreed and planned to move. Others said that La Conchita was their home. They wanted to be able to choose where they lived.

People who stay in La Conchita will always have a reminder of the day a mudslide killed 10 of their neighbors. All they have to do is look up at the hillside, now bare where a piece of the mountain is missing.

The mudslide left a deep scar on the landscape of La Conchita.

CHRONOLOGY

March 4, 1995	A major mudslide occurs in La Conchita, destroying four houses.
March 10, 1995	A smaller mudslide occurs near La Conchita. Retaining wall is built. Sensors are placed in mountain behind La Conchita.
December 27, 2004	Heavy rains begin falling on Southern California. They will continue for about two weeks.
January 10, 2005	A small mudslide closes the highway north of town in the morning. At approximately 1:30 P.M. a large mudslide devastates La Conchita. Rescue efforts begin.
January 11	Rescue efforts continue.
January 12	The last bodies are recovered from the mudslide. Governor Arnold Schwarzenegger declares Ventura County a disaster area.
January 13	The search for survivors ends. Residents are warned it is unsafe to return to the area.
January 22	Hundreds of mourners attend a memorial service for the victims.

OTHER DEADLY MUDSLIDES

1985	Mameyes, Puerto Rico—at least 129 people killed
1999	Villahermosa, Mexico—425 people killed
	Vargas, Venezuela—30,000 people killed
	Central America—1,900 people killed (mudslides caused by Hurricane Mitch and Casitas Volcano)
2002	Chuuk State, Micronesia—47 people killed; 109 injured
	East Africa—112 people killed
2003	North Sumatra, Indonesia—74 people killed; 200 missing; 100 injured
	Chittagong Hill Tracts, Bangaladesh—over 65 people killed
2004	Republic of Kazakhstan (Central Asia)—over 20 people killed
2005	Santiago, Chile—11 people killed; 200 injured
	Ankara, Turkey—17 people killed; 7 injured
	Guatemala—Hundreds left buried as whole towns are declared mass graves

FIND OUT MORE

Books

Green, Jen. *Earth.* Brookfield, Connecticut: Copper Beach, 1998.

Simon, Seymour. *Danger! Earthquakes.* New York: Seastar Books, 2002.

Ylvisaker, Anne. *Landslides* (Natural Disasters). Mankato, Minnesota: Capstone, 2003.

On the Internet

Federal Emergency Management Agency Disaster Preparedness for Kids
http://www.fema.gov/kids/

United States Geological Survey Kids Pages
http://interactive2.usgs.gov/learningweb/students/index.htm

American Red Cross: Landslide and Debris Flow (Mudslides)
http://www.redcross.org/services/disaster/0,1082,0_588_,00.html

Works Consulted

Seals, Brian. "La Conchita mudslide stirs memories of Love Creek catastrophe." *Santa Cruz Sentinel,* January 13, 2005.

"Calif. Mudslide Death Toll Rises to 10." *The Washington Post,* January 13, 2005.

"'I Had to Run for My Life.'" *The Washington Post,* January 12, 2005.

"La Conchita residents unsure whether to stay or walk away." *San Jose Mercury News,* January 29, 2005.

"Rescuers dig through rubble for survivors of La Conchita, Calif., mudslide." *Chicago Tribune,* January 13, 2005.

Hallman, Lesly C. Red Cross: "Mudslide Buries Small California Town," January 11, 2005. http://www.redcross.org/article/0,1072,0_312_3943,00.html

Kahn, Carrie. National Public Radio: "La Conchita Residents Stay Despite Known Risks," January 13, 2005. http://www.npr.org/templates/story/story.php?storyId=4282192

Okwu, Michael. MSNBC: "They were so pure and good." January 13, 2005. www.msnbc.msn.com/id/6821292/

CBS/Associated Press: "Lifesaver In Calif. Mudslide," February 9, 2005. www.cbsnews.com/stories/2005/01/11/national/main666278.shtml

CBS/AP: "Hunt For Mudslide Survivors Ends," January 14, 2005. www.cbsnews.com/stories/ 2005/01/14/national/main666949.shtml

CNN: "Rescuers search for missing in mudslide. Six dead, 13 unaccounted for in California." January 11, 2005. http://www.cnn.com/2005/WEATHER/01/11/california.mudslide/

CNN: "Search for mudslide survivors extended." January 12, 2005. http://www.cnn.com/2005/WEATHER/01/12/california.mudslide/

MSNBC: "Rescue called off in Calif. Mudslide; As town mourns, residents warned of persisting dangers." January 13, 2005. www.msnbc.msn.com/id/6780872/

United States Geological Survey: "January 2005 California Landslide," February 7, 2005. http://landslides.usgs.gov/html_files/landslides/05jan_ca/05jan_cafaq.html

GLOSSARY

chassis (CHAH-see)—the frame of the body of a car.

collapsed (kuh-LAPST)—fell down suddenly.

community (kuh-MYOO-nuh-tee)—a group of people who live in the same area.

debris (duh-BREE)—scattered pieces of something that has been broken or destroyed.

earthquake (URTH-kwake)—sudden, violent shaking of the earth caused by a shifting of the earth's crust.

erosion (ih-ROE-zhun)—the gradual wearing down of something like soil by wind or water.

geologists (jee-AH-luh-jists)—people who study the earth's layers of soil and rock.

microphones (MY-kruh-fones)—instruments that transmit sound.

retaining (rih-TAYN-ing)—holding something in or containing it.

sensors (SEN-surs)—instruments that can detect changes in heat, sound, or pressure and send that information to a controlling or warning device.

tons (TUNS)—a measure of weight; one ton equals 2,000 pounds.

vertebrae (VUR-tuh-bray)—small bones that make up a spine, or backbone.

INDEX